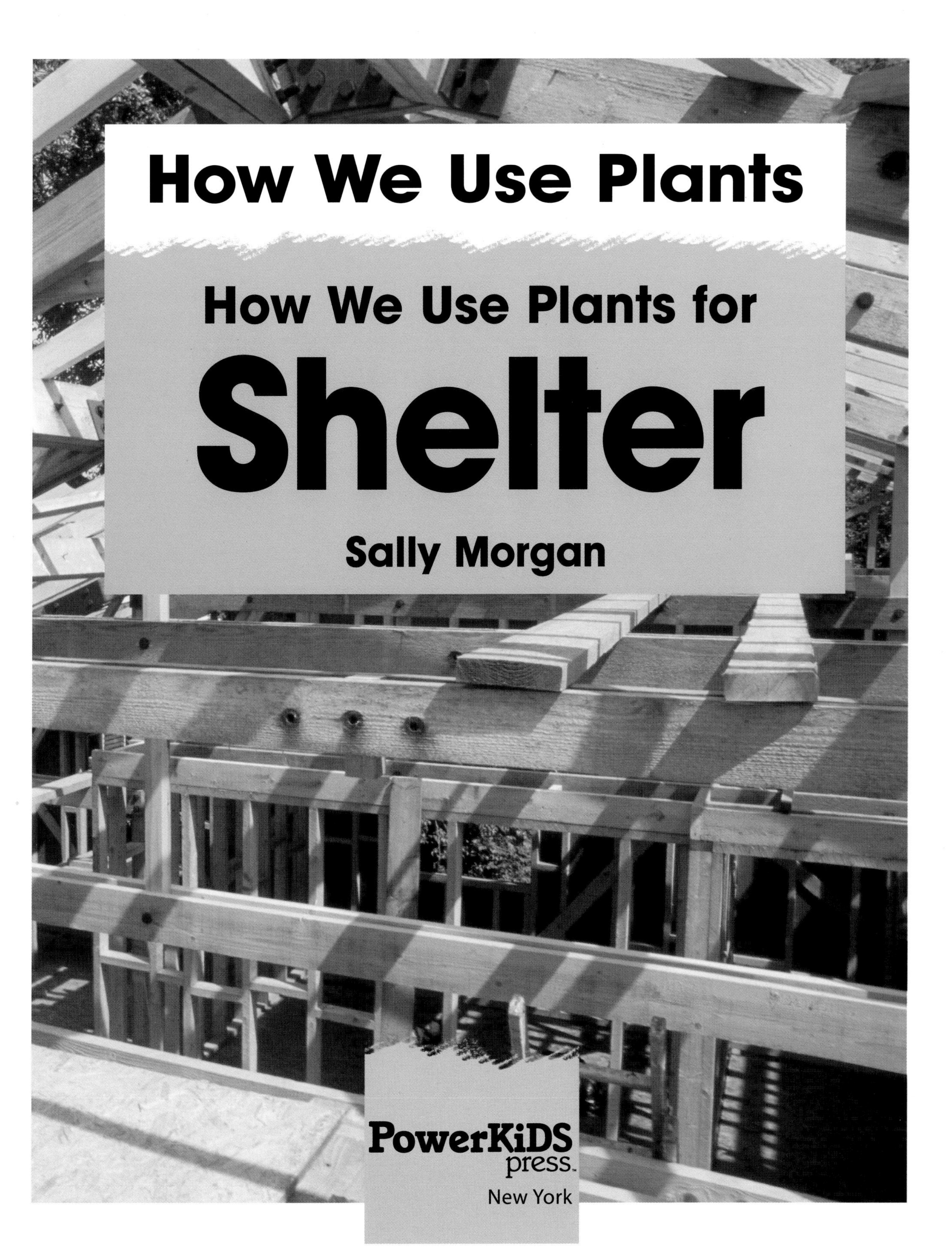

How We Use Plants

How We Use Plants for Shelter

Sally Morgan

PowerKiDS press.
New York

Published in 2009 by The Rosen Publishing Group Inc.
29 East 21st Street, New York, NY 10010

First Edition

Editor: Camilla Lloyd
Designer: Matthew Lilly
Picture Researcher: Sally Morgan

Picture Acknowledgments: The author and publisher would like to thank the following for allowing these pictures to be reproduced in this publication. Cover: Ecoscene. Ecoscene: 1, 4, 5, 6, 7, 10, 11, 12, 13, 14, 15, 16, 18, 19, 20, 21, 22, 23, 24, 25, 26, 27, 28; Najlah Feanny/Corbis: 8; Justin Kase/Alamy Images: 9; Philip Lewis/Alamy Images: 17; Steve Atkins/Alamy Images: 27 (bottom right).

With special thanks to Ecoscene.

Library of Congress Cataloging-in-Publication Data

Morgan, Sally.
How we use plants for shelter / Sally Morgan.
p. cm. — (How we use plants)
Includes index.
ISBN 978-1-4042-4422-1 (library binding)
ISBN 978-1-4358-2612-0 (paperback)
ISBN 978-1-4358-2626-7 (6-pack)
1. House construction. I. Title.
TH4815.M67 2008
690'.837—dc22

2007041597

Manufactured in China

Contents

Plants for shelter

Thousands of years ago, the first shelters made by people were made from plants. These shelters were very simple. The people used fallen tree **trunks** and branches to make a shelter like a **tent**. Then they covered their shelter with leaves to keep out the rain.

The roof of this house in Kenya is covered with palm leaves.

Today, modern homes use many types of building materials, such as bricks, steel, and concrete, but plant materials are still important.

Wood is used to make doors, window frames, floors, and the supports for the roof. In the tropical parts of the world, people use palms and bamboo for making their homes.

This cottage in England has a roof covered with reeds.

Plant materials should never run out, so long as new trees, palms, and other plants are planted to produce building materials for the future. The supplies of these materials are said to be **sustainable**.

Wood

Trees can be large plants. Each year, trees make new wood and their trunks get wider. Trees grow bigger and bigger as time goes on.

Your Turn!

- Take a look at the trees growing near your home. Are there any pine trees? Pine trees have needlelike leaves.

This man is using a chainsaw to cut down a large tree.

There are two main types of wood, **hardwood** and **softwood**. Hardwood comes from trees such as oak and beech. They are slow-growing trees with good quality wood. Softwood comes from fast-growing trees, such as pine and spruce. It is cheaper than hardwood and there is more of it.

The trunks of trees are transported on large trucks to the lumber mill.

Trees, such as pine and spruce, are planted in **plantations** as a **crop**. They are left to grow for 30 to 50 years and then cut down. The wood is taken to **lumber** mills and cut into planks.

Woody materials

Wood can be made into a variety of wood products. Plywood is made from three or more thin layers of wood that are glued together. This makes the sheets of plywood strong. Plywood is a cheap material, since it can be made from poor-quality wood. It is used on floors and for making furniture and cupboards.

Your Turn!

- Take a look around your home and see if you can find examples of plywood and particleboard. What were they used for? Can you find any veneers?

Tree trunks can be sawn into planks and used to make things.

Another cheap material is particleboard. It is made from wood chips glued together to form sheets. MDF stands for Medium Density Fiberboard. It is like particleboard, and it is made from fibers of birch and larch glued together.

The roof of this house is supported by wooden beams. The floor of the attic is made of MDF.

A veneer is a thin sheet of wood that can be stuck onto a poor-quality wood or particleboard to make it look like good-quality wood.

Wooden homes

When Europeans first arrived in North America, they made simple log cabins from wood. Trees were **felled** and the trunks were used to make the walls.

In Great Britain, many older houses that date back to Tudor and Elizabethan times have wood frames. The wood can be seen from the outside on this house.

Larger structures, such as houses and barns, can be built using a wood frame. In a wood-framed building, long planks of wood form the framework and they support the whole house. The planks of wood are held together with wooden pegs.

Did You Know?

The wood used in wood frames is often called green wood, because it is freshly cut and it has not dried out. It gradually shrinks as time goes on.

This carpenter is building the wood frame of a new house.

The walls are formed by filling the spaces in the wood frame with brick or smaller planks of wood. Sloping planks of timber are used to build the roof. Then tiles, slates, or **thatch** are laid on these roof supports to keep water out.

Thatching

For thousands of years, people have used reeds and **straw** as roofing materials. A roof made of plant material is called a thatched roof. Plants were chosen to put on the roof because they were light in weight. The buildings of the time were not strong enough to support the weight of anything heavier.

Did You Know?

The first Globe Theatre in London, England, was built in 1599. It burned down fourteen years later, because a spark from a canon used on the stage landed on the thatched roof and started a fire.

Reeds are cut and bundled together, ready to be used on a roof.

Many types of plants can be used as thatch. Palm leaves are used in tropical countries. In Europe, reeds, straw, willow, and grasses are used.

Thatched roofs are common on the island of Madiera, which lies in the Atlantic Ocean.

The plants are harvested and made into small bundles. The bundles are laid on the roof and secured in place with twigs. A second layer is added to make sure the roof keeps water out. The thatcher can carve interesting shapes into the roof.

The walls and roof of this Sri Lankan family's house are made from palm leaves.

Straw homes

Straw is the dry **stalk** of a cereal plant, such as wheat and barley, that is left behind after the seeds have been harvested. Most straw is used as animal bedding, but it can be used as a building material, too.

Did You Know?

The first straw bale buildings were built in North America during the late 1800s. People living on the plains of Nebraska did not have any trees or stone to make their homes, so they used leftover bales of straw.

Straw is left on the field by the combine harvester after the seeds of the cereal have been collected.

Houses can be made from straw bales instead of bricks. Straw bales are good at keeping out the rain and cold, and keeping the heat inside the house. They are strong but lightweight.

Beams of wood or steel stop the bales moving. The outermost bales are covered with mesh to keep rats, mice, and birds from nesting in the straw!

Straw is allowed to dry and then bundled into bales. The bales are stored in a dry place until they are needed.

The walls of this new house are made from straw bales on a wood frame. Holes are cut to make windows and doors.

Tents

Tents are temporary shelters that protect people from wind, rain, and the cold. **Nomadic** people move from place to place, often following their herds of grazing animals. They use tents because they can be put up and taken down easily.

These weavers in the Himalayan Mountains are working in a tent made from plant materials.

Canvas tents are ideal for camping, because they are small enough to be carried and they are quick to put up.

Tents vary in size. The smallest ones are for one or two people, but the largest are the size of many rooms. Tents can be supported by a central wooden pole, a series of hoops, or by several upright poles.

Once the frame is in place, material is laid over the frame and secured in place so that it does not blow away. Often, this material is **canvas** made from the cotton, jute, or hemp plant.

Did You Know?

One of the largest tents in the world was the Grand Marquee at the Chelsea Flower Show in London, England. It was used during the 1990s and covered 3.5 acres (1.4 hectares) and weighed 65 tons.

Bamboo for building

Bamboo is an important building material, especially in Asia. It is used in the same way as wood. Bamboo is a fast-growing grass. Each year, the canes or **stems** are harvested and new ones grow back the following year.

The scaffolding used on this building in China is made from bamboo.

The largest bamboo canes are used to make the frame of a house and even to support the roof. Bamboo canes can also be used as **scaffolding**, in the same way that steel is used.

Some of the longest bamboo canes are 10 feet (3 meters) long and reach a diameter (width) of 5 inches (12 centimeters).

Bamboo canes are made into boards that can be used to make walls and floor coverings. The canes are split down the middle and woven together. Glue is spread on one side to harden the whole structure, so that it forms a stiff board.

Bamboo canes are hollow. Can you think of any uses for hollow canes?

Paints

A paint has three main ingredients. It contains a **pigment** that gives the color, a **binder** that sticks the pigment to the object being painted, and a **solvent** that holds the pigment and binder together. Without a solvent, the pigment and binder would form two separate layers in the paint.

In watercolor paints, the solvent is gum arabic, something that comes from acacia trees.

Paints are used to add color to walls, wood, and other surfaces. Paints can protect wood, too. Wood that is left out in the rain will eventually rot and have to be replaced. Paint keeps most of the water out, so the wood lasts longer.

All these ingredients can come from plants, although many modern paints use **artificial** ingredients. Binders can be made from gum or plant sap. A natural solvent is linseed oil. It can also be made from the peel of citrus fruits.

It's A Question!

Decorators use many types of paint, one kind is called gloss paint and another is matte paint. Can you find out which type of paint would be used to paint on wood?

Unused paints should be taken to special sites, such as this one in North Wales. Then it can be disposed of safely or sent to be recycled.

Plant pigments

Many different pigments can be found in plants. For example, you may get a green stain on pale-colored clothes if you sit on the grass.

The stains from orange and grape drinks are really difficult to get out of clothes. This is because the pigment sticks to the cloth and cannot be removed.

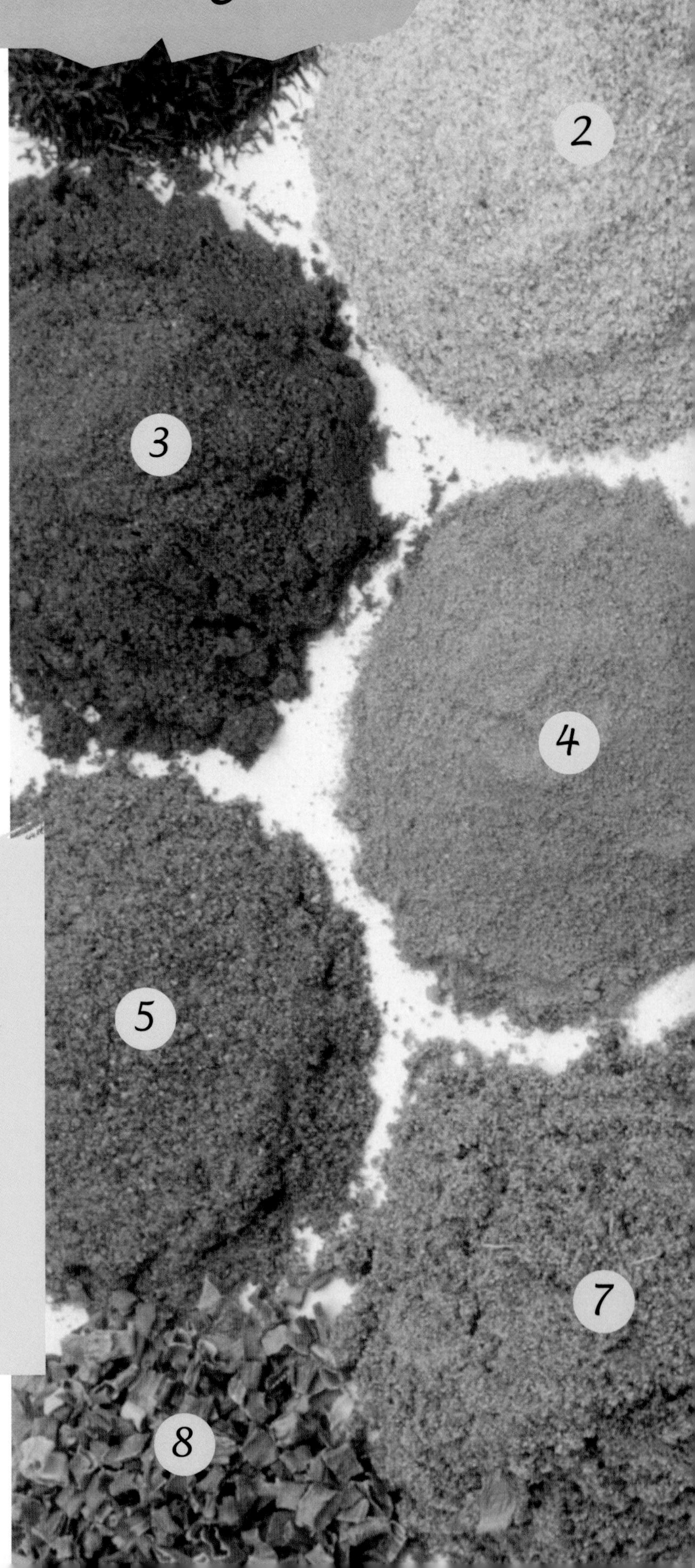

These herbs and spices contain different-colored pigments.

1. Dill
2. Coriander
3. Paprika
4. Turmeric
5. Cinammon
6. Nutmeg
7. Cumin
8. Chives
9. Rosemary

Some of the oldest plant pigments in use include saffron, turmeric, and carrots. However, there are many other pigments. Common examples include walnut, blueberry, beet, paprika, and hibiscus flowers.

Your Turn!

- It is easy to see how plant pigment can be used to color something. You need a piece of white cotton fabric and some chopped beet. Place the beet in a pan of water and simmer gently. Add the fabric and leave for about 5 minutes. Remove it using a wooden spoon and allow it to dry. You will need an adult to help you. It should now be purple from the beet.

A purple-red pigment is obtained from the hibiscus flower.

Plants for walls and floors

A number of plant materials are used in the home. Curtains are made from cotton fabric, and chairs are made from rattan. Paper wallpaper made from trees can cover the walls.

The insulation in this new home comes from old newspapers that have been broken up to form a paste. Paper is made from trees.

In some new homes, the **insulation** in the walls is made from recycled paper. Paper insulation is full of air spaces, and this traps the heat inside the home so that little escapes.

Floors may be covered by floorboards made of wood. Carpets are laid over the floorboards. Although many carpets are made from wool, plant fibers, such as sisal, sea grass, and hemp, can be used, too. These carpets are very hard-wearing.

This man is making a woven mat from large palm leaves.

Your Turn!

- Is the doormat at the entrance to your home made from plant fibers? Take a look. Mats made from sisal or coconut fibers are very rough, so that they trap the dirt from the bottom of shoes.

How the oak tree is used

Oak trees have been used as a source of high-quality timber for thousands of years. An oak tree can grow to a great size, and the timber is ideal for building ships and homes as well as many other things.

Timber:
Oak wood is a long-lasting hardwood. It is used to make wood-framed buildings, furniture, doors, and windows.

Charcoal:
The smaller branches from an oak tree are not suitable for timber but can be made into **charcoal**. Charcoal is used as a source of heat on a fire or barbecue.

Acorns:
The fruit of the oak tree is called an acorn. Acorns are eaten by many animals, such as squirrels and pigs.

Furniture:
Furniture, such as beds, tables, and chairs, is often made from oak. The oak can be stained to give a light or dark color to the wood.

Build your own model hut!

Take a look at the photo on this page. This is the first stage in the building of a traditional stilted hut in Zambia. See if you can build a model of this at home using plant materials. You will need a piece of cardboard about 8 x 6 in. (20 x 15 cm), some twigs, dried grass or leaves, glue, and some string.

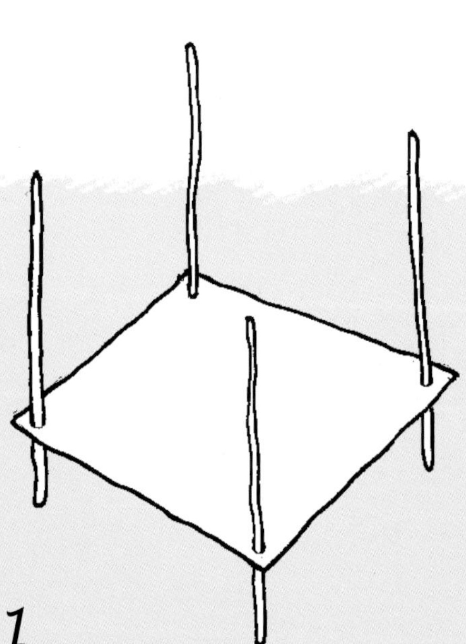

Step 1

The piece of cardboard forms the floor of the hut. Make a small hole near each of the four corners. Take four twigs that are about 8 in. (20 cm) long and push one through each of the holes. Each twig needs to stick out about 2 in. (5 cm) below the cardboard.

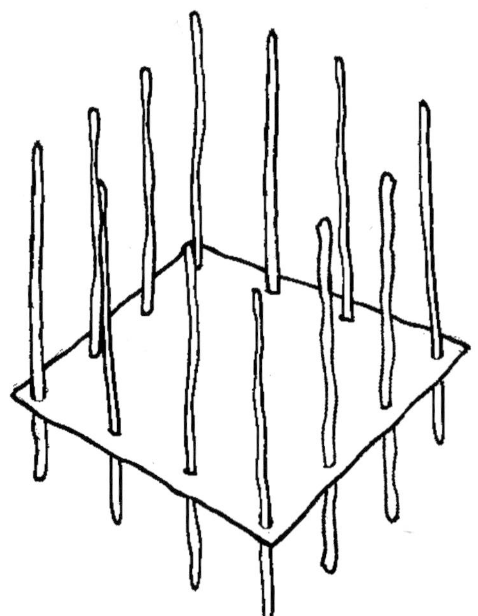

Step 2

Make two holes along each side and push some more twigs through. Now your model is at the same stage as the hut in the photograph.

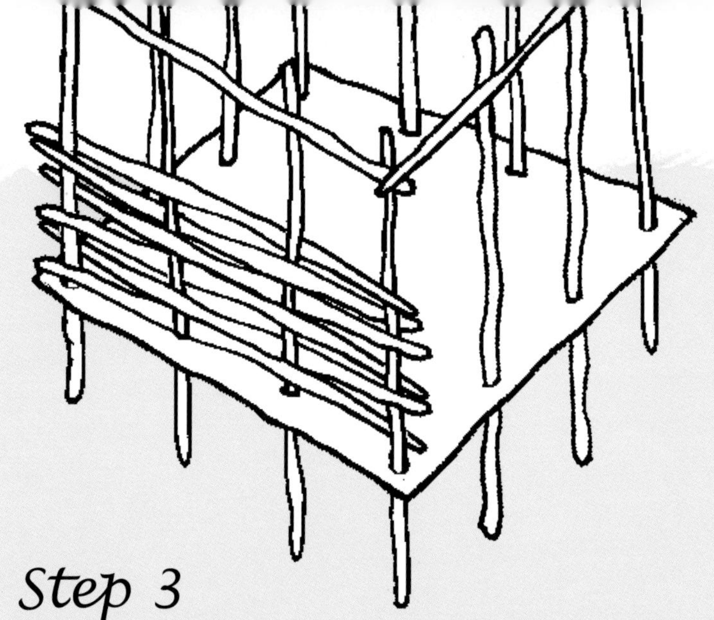

Step 3

Your next job is to thread some long twigs horizontally through the top of the vertical twigs and secure with string. These twigs form the top of the four "walls" of the hut. You can weave some dried grass or twigs between the uprights to form a woven wall.

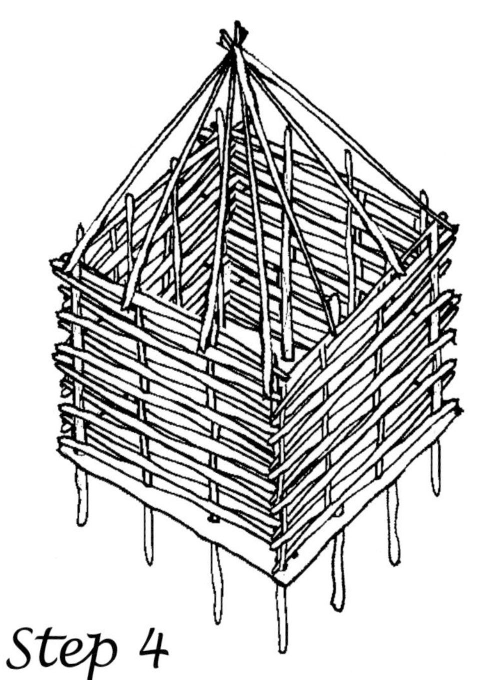

Step 4

The roof can be made from twigs attached at an angle to the top of the wall. You will need quite a few roof supports to attach the thatch securely with glue.

Step 5

The thatch can be made from dried grass or leaves and tied with string. Cut out some doors and windows. You can make the doors from wood or from a woven rectangle of grass.

Glossary

artificial not natural, made by people.

binder the part of paint that makes it stick to the object being painted.

canvas a rough cloth used to make sails and tents.

charcoal the black coals formed by burning wood slowly. Charcoal can be used to heat things on a barbecue.

crop a plant grown by people for a special use.

felled cut down.

hardwood wood that comes from slow-growing trees.

insulation something that prevents heat escaping.

nomadic nomads are wandering people, nomadic is a characteristic of wandering people.

pigment a substance that gives color to paint.

plantation an area of land where crops or trees are planted.

scaffolding a series of poles and platforms put up around a building, so that workmen can reach the higher parts of the building.

softwood wood that comes from fast-growing trees.

solvent the part in paint that holds the pigment and the binder together.

stalk stem of a plant.

stem part of a plant that supports the leaves.

straw the dried stalk of a cereal plant.

sustainable a way of managing a resource, such as wood from trees, so that there is enough for the future.

tent temporary shelters made from material such as canvas and supported by poles.

thatch plant materials used to make the roof of a house.

timber (or lumber) wood from trees that is mostly used for building materials.

trunk the name given to the stem of a tree.

wood a substance made by trees that makes up much of the tree trunk.

Further information

Books

Looking at Plants: Plants and People by Sally Morgan, Chrysalis Children's Books, 2004

Looking at Plants: Roots, Stems and Leaves by Sally Morgan, Chrysalis Education, 2002

Plants and Us by Angela Royston, Heinemann, 2001

Web Sites

Due to the changing nature of Internet links, PowerKids Press has developed an online list of Web sites related to the subject of this book. This site is regularly updated. Please use this link to access this list: www.powerkidslinks.com/plant/shelter

Question answers

P18 pipes

P21 gloss paint is an oil-based paint and is used on wood. Matte paint is used on walls.

Index